"You are Unique"
Aren't you?

"You are Unique" Aren't you?

Kumar Vuppala

authorHOUSE®

AuthorHouse™
1663 Liberty Drive
Bloomington, IN 47403
www.authorhouse.com
Phone: 1-800-839-8640

Published by AuthorHouse   12/19/2012

ISBN: 978-1-4772-4250-6 (sc)
ISBN: 978-1-4772-4251-3 (e)

CONTENTS

PERSONAL BRANDING

<u>ARTICLES</u>

"SOCIAL MEDIA"

PERSONAL BRANDING

Kumar Vuppala

Biologically, it is quite clear that everyone is

Unique, right from birth.

We are told that the human genome is a string of some 70 billion pairs of protein molecules arranged in different sequences. Of course, as I understand it, many parts of the sequence are more or less fixed, but there is still a lot of room for a seemingly infinite number of combinations and permutations that determine individual characteristics.

If ordinary implies pretty much the same as everyone else, biologically, we are not.

CHAPTER 1

THE
IMPORTANCE
OF BUILDING
YOUR OWN
PERSONAL
BRAND

It's not only celebrities who have their own personal brand. As the Web gets increasingly more social, it's essential for everyone to take branding into consideration. Just like products and services, your personal brand tells everybody about your unique characteristics. Here's why it's so important to consider.

BRANDING BUILDS TRUST

When you've successfully built a personal brand, your customers and fans will better understand why you do what you do. This engenders trust and credibility, which are both essential. When people know and trust you, they're more comfortable dealing with you and more likely to buy your products or services.

THE REAL YOU

On social media, everybody loves people who are 'real.' They're not into the sales talk or corporate image. Branding tells people not just what you do but what you stand for, which is even more important. Your brand vision is what stays in people's minds, not the specific goods that you offer.

PERCEIVED VALUE

People respond emotionally to brands, much more so than to products or benefits. When you're well-branded, you have a name in your industry. You're seen as a key player and this raises your value in your customers' eyes. This also means you can charge more for your services.

YOUR EDGE

Your personal brand helps you beat the competition by showing the unique things that you offer. The competition may win in other areas, but they can't compete with your unique natural qualities. Branding helps you carve out your own special niche in the marketplace.

IMAGE CONTROL

You've got a personal brand whether you realize it or not. Your personal brand is your reputation. Are you in control of it? When you handle your own branding, you can take control. This allows you to better deal with attacks or negative comments about you.

BUILD A BUZZ

If you brand well, you can get a buzz going. Your name gets passed around by your happy customers and fans, and this means that you get more business through referrals. Branding has a kind of snowball effect if you do it right.

JOB SECURITY

Job security is almost non-existent today and your personal brand protects you against economic problems. Even if you work for a big corporation, you're a company of one. And, just like a company of many, your branding keeps you alive. If you're a freelancer, it's a key element to survival.

How do you get started building your own brand? A good place to start is to look at other successful personal brands in your industry. Follow them on social media or check out their websites. What sets them apart? What's their brand message? How do you feel about them? Look at all of the different ways they create their personal brand and take lessons from them. You can apply these techniques to building your own personal brand.

CHAPTER
2

The rise of social media has made it more important than ever to have your own personal brand. Like a product brand, your personal brand tells people who you are and what value you bring to the world. Your personal brand is your reputation and it's essential to take control of it, whether you're a growing business or a job hunter. There are 5 key characteristics that it needs to have.

CONSISTENT

All of your actions should be consistent with your brand image. You'll become known for something that you always do and you shouldn't stray too far from it. When you're consistent, it shows that you're reliable, and this helps to establish your credibility. When you're inconsistent, this creates a disconnect that's not conducive to branding. Although you should be consistent, remember that your personal brand isn't static. It should evolve over time.

UNIQUE

Uniqueness is key because your brand needs to set you apart from everybody else. However, you don't want to force uniqueness by trying to be something that you're not. Instead, look for your natural qualities that are unique to you. Don't create it; discover it. What is your passion? What are you particularly good at? What makes you different? Look for the things you don't see in others.

PERSONAL

The personality of a business brand is more important to its fans than its actual products or services. This is no different for your personal brand. Show your personality as much as possible, especially when interacting with people online. Most people want to be seen as a friendly, generous, helpful expert. Spend your time connecting with others and learning from them.

MEMORABLE

Memorability is probably the most important characteristic needed for personal branding. Even if you're the world's greatest blogger, if no one remembers your name 10 seconds later, your branding has failed. Highlight a small number of attributes or skills and don't try to be master at everything. These few things that you really excel at will become linked to you in the minds of others. Be brazenly different and go against the grain. It will make you a leader rather than a follower.

CLEAR

Make sure that everything you produce shouts out loud what it is you do, what value you offer, and how you're unique. Don't make people think about it and don't be subtle. Tell the world what it needs to know about you in as few carefully-chosen words as possible.

Personal branding is a relatively new concept and there isn't a textbook that tells you how to do it. The best way to work on your brand is to watch and learn. Identify those who brand themselves well and follow them. Study what they do and learn to do it yourself. Pay especially close attention to how their words and actions are consistent, clear, unique, personal and memorable.

CHAPTER
3

Social media profiles are essential in personal branding today. They tell people all about you and the value you offer, and they give your fans a place to talk and get involved. This tool makes branding easy, but there are a few guidelines that you need to follow.

YOUR PROFILE IMAGE

Your picture is kind of like your logo. Everybody will instantly recognize your brand when they see it. Make it something that looks natural. A headshot that shows part of the neck and shoulders is always good. It shouldn't be too big or small. Use the same photo for all of your profiles. If you have others that you'd like to use, put them in your albums. Try not to change your profile picture and if you do, don't do it often.

Logos really aren't appropriate for a personal profile. Remember that it's you and not your company that you should be branding. If it's a Facebook fan page or a brand page on LinkedIn, logos are fine, but for your personal profile, stick with a picture.

CONSISTENCY IS KEY

Most social media sites allow you to customize your profile. You should by all means do this. Use the same design patterns and colors that you use on your website. A great way to do this is to take the horizontal banner of your site and make a vertical version for your profiles.

Consistency is also very important with your profile information. Each profile should have the same basic information written in the same general tone. This makes you appear reliable and trustworthy. Some sites like Twitter don't allow you as many words in your profile, so create a shortened version with just the most essential information.

Never leave a profile incomplete. Fill out everything. If it's incomplete, this means that you don't have much to say about yourself. It may also make you look lazy. You should be giving people as much information as possible.

Connect all of your social media sites as well as your website. This helps people understand that it's you they're connected with. It affirms that you're the brand they know and love. It also offers more touch points for you and your customers.

WHAT TO WRITE

Your social media profiles should be consistent, but there's one area where you should mix it up, and that's content. Each social media site has certain types of content that it likes. Twitter and Facebook like off-the-cuff comments and personal updates. Networks like LinkedIn and Google+ are more professionally oriented. People on those sites like news stories and helpful tips. It's okay to cross-post your posts, but it's better if you can customize.

It's also nice to offer something different on each social media site because many of your fans will be the same everywhere.

Whenever you're using social media, you should always be on-brand. Be mindful of what you do because you never know who will see it. Follow all social media etiquette and keep things positive. You should show your unique personality, but always consider how your actions will look to your fans. Maintain a high level of professionalism.

CHAPTER 4

Branding is important for anyone who's online, and writing your own personal blog is a perfect way to brand yourself. You need a personal blog if you want to tell the world about your unique talents and the value that you bring. Here are a few reasons why a personal blog is so important:

- **It's Your Social Resume.** Your blog is an online resume that shows people what you can do and what value you bring. It shows off your strengths.

- **It Keeps You Relevant.** In order to establish an effective personal brand, you need to establish your expertise. Your blog shows that you're up-to-date on what's going on in your industry.

- **It Builds Trust.** One purpose of branding is to build trust. People want to know who you are before they buy from you or hire you. Your consistency in your blog posts helps to build this trust.

- **It Gives You an Avenue of Communication.** Blogging is all about networking, communicating and sharing interests with like-minded people. You can build relationships with other bloggers and with your readers.

- **Image Control.** Your blog is your 'home on the Web.' It can help you manage your reputation.

- **You Can Tell Your Story.** A compelling story is important in branding. It shows who you are and where you came from. Your blog gives you a chance to tell yours.

Naming Your Blog

The best name for your personal blog is simply your own name. However, there are a few things to keep in mind. For one thing, if you use your name for your personal blog, you should be the only person writing content. It's okay to occasionally accept guest blog posts, but the purpose of the blog is to sell you, not other writers.

You shouldn't use your own name if the purpose of your blog is to sell products or services. For example, if you're going to do affiliate marketing, use something else for the name. Your personal blog should focus only on you. If you're using it to sell unrelated products, this will cause a disconnect for your readers.

One more thing to keep in mind is that you should only use your name if you plan to blog long-term. If you think you may someday sell the blog, use something else. It's nearly impossible to sell a website named with a person's name.

What Should You Write?

When writing your personal blog, focus not on what you want to talk about, but what your readers want to hear. What would help them the most? Answer their questions and offer tips to help them with their problems. You might also post news about your industry or other topics that you know about.

Try to focus your blog posts on getting some dialog going. Ask questions or present arguments and encourage your readers to voice their opinions. You want to get as many people commenting as possible.

Don't forget that the personal touch is what really sets a brand apart. This means that it's alright to sometimes write purely personal posts. You don't want your blog to be all about what you did today or how you feel about something you saw on TV, but it's good to let your human side come out. Your readers want your helpful advice, but they also like to know more about you as well.

CHAPTER
5

You know that you need a strong personal brand to succeed today, but how do you go about establishing one? Personal branding is a long-term strategy that pays off down the road, but you have to start somewhere. Here are 8 ideas on how to creatively set yourself apart from the herd.

BLOG

Writing a blog is the absolute best way to brand yourself online. It allows you to share your expertise, show your personality, and get your fans involved. These are all essentials for branding yourself and a blog lets you kill all of those birds with one stone.

SIGNATURE WORDS AND PHRASES

Sit down and try to write a short sentence or simple phrase that communicates to people exactly what you do and how you benefit them. This is what's known in the marketing world as a 'tagline.' For example, if you're an SEO expert, you might write something like, 'I get you found online.' It serves both purposes—telling what you do and emphasizing benefits.

YOUR BRAND PHOTO

Like a logo, a brand photo instantly identifies you. It also gives people a good first impression which might be the only one they get. Make sure that you look good in the picture and that you're smiling if you want to convey a friendly image. Be consistent with your image, using it everywhere you go. You might consider making a few different versions; for example, a more professional suit-and-tie picture for LinkedIn and something more relaxed for Facebook.

REGULAR EVENTS THAT YOU RUN

Get involved and start hosting regular events. Events show people that you're passionate about what you do and it's a great way to get them involved directly in your brand. Events are also a wonderful way to network. Online you can do podcasts and webinars. Both are cheap options that offer global reach.

SPEAK IN FRONT OF GROUPS

Look around for opportunities to speak in front of groups. When you're a recognized speaker, this boosts your credibility and gives you instant expert status. Like hosting events, it also gives you a chance to network. If you're not a great public speaker, don't worry; a little practice will turn you into a confident speaker quickly.

MENTOR OTHERS

Offer coaching and mentoring, even if you don't get paid well for it. This adds immensely to peoples' perception of you as an expert. If you can teach others about your subject matter, this shows that you really know your stuff.

CREATE AN INFOGRAPHIC

Infographics are all the rage now and you can create one that's all about you. Your own personal brand infographic is like a visual resume. There are online tools that you can use to create your own infographic cheaply and easily.

A COMPELLING STORY

People love a good story and it's an important component to your branding. For example, you're a city kid who grew up poor and then discovered Internet Marketing. Or, you're a stay-at-home mom who does freelance writing. A unique story about how you got here makes your brand more memorable and helps you to stand out.

Remember that the purpose of all of your branding efforts is to set yourself apart from everybody else. Use these creative methods to show how you're different and the unique value you bring to what you do. Branding really takes off when your personal passion comes through, so let it shine.

CHAPTER 6

An easy way to think about branding is that it's your reputation. You're consistent, you're an expert, and it's clear what you can do for people—This is what your brand tells people about you. So it's very important to think about branding in everything you do, including the emails you send to your list. Your emails are your touch points with your subscribers and they need to be on-brand.

STYLE AND DESIGN CONSIDERATIONS

Choose templates, colors and visual styles that reflect your business's image. This is not a tiny, unimportant detail that you can afford to neglect. Your email design features should be in line with all of your other materials. If you make offers through your email messages that are also available through your site, they should be the same in design for both your website and email.

Don't use a standard template and whatever you do, don't use a standard 'thank you' page. The 'thank you' page is what they see when they first subscribe to your list and it gives a first impression. It should be designed along the same lines as all of your other materials.

Keep the design of your emails simple. If you need to, simplify the design of your website to suit your emails.

Simpler design offers a better user experience. For most people, emails that are cluttered look like spam. Keeping it simple also makes it clearer what you're offering them.

YOUR EMAIL CONTENT

The tone of your content is even more important. Is your brand youthful and casual, or businesslike and professional? Are you tough and no-nonsense? Are you brainy and geeky? The wording of your emails needs to convey this image. If there's a disconnect between your wording and what you're offering, it won't build the trust you need.

Greetings, signatures and PS's can be dashed off quickly, but you should take some time to consider the tone here as well. The greeting should be personalized and include the list member's name. Never send out a message that begins with a generic 'Greetings' or 'Hello.' The signature may seem insignificant, but don't forget to include it. Leaving it off makes the message seem abrupt and sloppy.

OTHER CONSIDERATIONS

Always read through the message at least once and check for grammar and spelling mistakes. Run it through spell check as well. No matter what image you're going for, mistakes in your emails aren't going to help you brand yourself as anything except an amateur.

Your logo should be displayed in your emails (if HTML) along with icons for your Facebook, Twitter and other social media sites. Make sure that all of these are clear and readable.

When you're writing and editing email messages, always look at them from your recipient's point of view and not your own. Put yourself in their place. Send yourself a test email before broadcasting a message to see how it looks. Remember that customers are more likely to respond when they recognize your brand and it speaks strongly to them.

CHAPTER

7

Creating a branded resume makes your job search much easier. This is a resume that clearly sets you aside from others and sells them on your unique qualities. Today, it's not a matter of 'Why should I hire you,' but 'Why should I hire you instead of all these other people?' Your branded resume should tell them that.

LASER TARGETING

First of all, who are you targeting and how well do you know them? Creating a branded resume that speaks directly to the hiring manager and their needs is what gets you the job. Get to know the industry and find out exactly what employers are looking for. In your resume, use industry-specific jargon and terminology so that you're speaking their language. On the Internet, research job listings and companies to get ideas. If possible, talk to someone in the industry to get advice.

EMPHASIZE YOUR NATURAL STRENGTHS

Your natural strengths are the things you can do without even trying. They come through not so much in your hard skills, but in your soft skills. These are things like your ability to problem solve, communicate, work with

21

others and innovate. Take a good look at your personality and your work experience in order to discover these natural strengths, and make sure that they're clearly communicated on your resume.

BECOME A HIRING MANAGER

Look at your resume from the point of view of a hiring manager. Does it clearly tell them what benefits you're going to bring to their company? Does it spell out loud and clear the value you bring that others lack? Branding is about more than just logos and design. Good branding shows your unique contribution.

VALUE PROPOSITION

Create your own unique value proposition. This is one sentence that says in as few words as possible exactly what benefits you bring. Trim any words or phrases that don't specifically explain these benefits. Make sure that your value proposition is clear from the very beginning of the resume to the end.

PROVIDE EVIDENCE

Make bold claims about what you can do and then use your resume's sections to back them up. Your job experience and educational background sections shouldn't just be laundry lists of job responsibilities. They should include detailed facts and figures that emphasize the results you've gotten. To a hiring manager, this translates to the results you'll get for them as well. Spend some time brainstorming your past achievements, focusing on how you've helped employers, customers, coworkers and clients.

QUICK AND FOCUSED

After you create your resume, trim it of anything that doesn't contribute to branding you. It should be simple and easy to read. The most important bits should be at the top and the left-hand side because these are the areas where the eye naturally falls.

When you're writing your resume, remember that it will provide talking points for your interview. Include all of the success stories that you'd like to tell the hiring manager. Make sure that everything contributes to creating the personal brand that is You.

THE STEPS IN
SETTING UP
YOUR OWN
PERSONAL
BRAND
SYSTEM

Branding needs to be consistent and that's why you need to have a system in place. This includes all of the things you need to do before you start branding as well as routine tasks you'll continue to perform ongoing. Everything needs to be as efficient as possible. This is how the branding pros do it and you can do it as well.

BE CLEAR FROM THE BEGINNING

Write down a statement about your brand's vision. It should be as short and concise as possible, and should include who you are, what you do, and your promise to your audience. Whether or not you use this as an official statement doesn't matter; what matters here is just getting it clear for yourself. What do you want people to remember you for? What's your unique position in the market?

IDENTIFY YOUR NATURAL STRENGTHS

Your natural strengths provide the engine for your personal brand. It's these strengths that will shine through everything you do. What are you good at naturally without even trying? Discover these strengths and figure out ways to make them work for you. When you know what you do well, you don't need to be egocentric or fake. Your promises will be real.

CHOOSE YOUR TOOLS

Branding is about getting your message in front of the people who need to hear it. The Internet gives you lots of avenues for doing this. These include Twitter, Facebook, YouTube, LinkedIn, Google+, your own blog, article directories, forums, press release sites, podcasts directories, and more. Brainstorm a list of these avenues for exposure and decide which you're going to use. You should also devote time to searching out new sites and strategies.

CREATE YOUR DESIGN

Your design features should be consistent. Choose fonts, colors, phrases and images that embody your brand. Take a good headshot that reflects the image you want to convey. Make sure it's not too close or too far away. Choose other images and templates that you'll use. Look at some sites and profiles of people that have strong brands to get ideas.

CREATE A SCHEDULE

Take all of your routine branding tasks and budget and schedule them. Decide how much time you're going to devote each day to each task. If you're not sure what these tasks should be, start with your goal (the statement you wrote to clarify). How can you achieve that? Keep asking yourself 'how' until you arrive at concrete tasks that will get you closer to it.

LISTEN TO PEOPLE

Learn about your audience by getting active on social media and subscribing to like-minded blogs. See what people are sharing, bookmarking and talking about. The better you know your audience, the better you'll be able to tailor your offers to them. You should also sign up for alerts so that you'll know whenever your brand is mentioned. Keep a careful eye on what people are saying about you.

GET ORGANIZED

Put all of your tools and resources in one place. Social bookmarking sites are great for this. When you have everything organized, it saves you lots of time and makes your personal branding much more efficient.

The most important thing is to monitor everything you do and look for results. If certain daily tasks aren't getting you results even after a few weeks, drop them and focus on something else. Whenever possible, split test your images, posts, and offers to see which version your audience prefers. Finally, because things are always changing in the online world, devote some time to searching out new ideas. Stay creative!

CHAPTER
9

SHOULD YOU OUTSOURCE YOUR PERSONAL BRANDING?

It might come as a surprise to you to discover that many people outsource their personal branding. How can you do this when your personal brand is supposed to be the 'real' you? While it can be risky to outsource branding, there are ways to do it right. You can find companies that specialize in turning your brand vision into tangible results.

THE RISKS OF OUTSOURCING YOUR PERSONAL BRAND

Probably the biggest risk of outsourcing your branding is letting the cat out of the bag. If your audience discovers that it's not you behind the social media profile, this can damage your credibility considerably, and credibility is the cornerstone of a good brand. Especially if you offer PR or branding services, they'll wonder why you didn't do it yourself.

The other risk is that you may hire a company that doesn't deliver. There are many companies that offer branding and reputation management services and you need to make sure that you find one that's good at what they do. Their job is two-fold:

1) To effectively communicate a brand image that's memorable, unique and consistent.
2) To make it truly match who you are so that there's no disconnect.

One more thing to consider before you outsource your personal branding is that it's not cheap. Quality branding services can be quite costly.

TIPS ON OUTSOURCING

Some people outsource their personal branding without any problem whatsoever. They can do this because they don't outsource their entire brand creation from the start. Instead, they outsource certain parts of their continued branding efforts.

You can start outsourcing your branding once your it's established and well-defined. The vision should come from you. What the branding specialists are actually doing is continuing and expanding your branding, not creating it from scratch.

A good guideline is to ask yourself with each task, 'Will this hurt my brand if they mess it up?' Of course, you should do your best to find a company that *won't* mess *anything* up; but consider the damage to your reputation if they do and don't outsource tasks that are risky. For example, you might hire a company to design your social media profiles for you, but you should probably do the actual posting of comments yourself. If your social media design looks a little off, you can fix it. If one badly phrased comment shows all of your fans that it's not you commenting, you'll lose lots of trust.

Transparency is another thing that helps. If you have people helping you with content creation or updates, let your audience see that. Give them their own profiles and allow them admin access to your blog, site or social media profile. This maintains your fans' trust while also getting you the help you need.

You can minimize potential damage by carefully overseeing everything your virtual help does. Have them send the posts to you for editing and tweaking. You can alter any language that seems funny and maintain control of everything before it goes out.

FINDING THE RIGHT HELP

When you're looking for a branding company to outsource work to, check them out carefully. Check references, read their client lists, and look at their portfolios. Speak to a live person before setting up the job and make sure that you can maintain control over everything they do. They should be responsive and reply to you quickly. Your personal branding is extremely precious and you should be extra careful about who you hire.

CHAPTER
10

TOOLS AND RESOURCES

PART A: INSPIRATION

WORKSHEET:
CREATING
YOUR
PERSONAL
BRAND

1. People you admire in your industry:

2. What are the characteristics of their personal brand? (e.g., background story, signature phrase, nickname, style, etc.)

3. Where are they seen most often? (Online sites, offline, seminars, etc.)

4. What do you admire about them?

5. What makes them unique?

PART B. BRAINSTORM YOUR BRAND

1. Who are the people you want to appeal to? (Characteristics)

2. How do other people see you now?

3. What do you want to be known as? (e.g., the person who does X)

4. What's your story? (e.g., why you do what you do)
5. What's your style? (e.g., casual, professional, grunge, etc.)

6. The skills you want people to know you have (natural, valuable talents)

7. Signature phrase or value proposition (just a few words):

8. What's unique about you?

DEFINE YOUR BRAND

1. Your story

2. Personal Values

3. Personal brand statement (state who you are in 1 sentence)

4. Your unique value proposition (how you uniquely provide value to your market)

5. Mission (why you do what you do)

CHAPTER
11

PART A: INSPIRATION

INSTRUCTIONS
FOR
COMPLETING
THE PERSONAL
BRANDING
WORKSHEET

1. WHO ARE THE PEOPLE YOU ADMIRE MOST IN YOUR INDUSTRY?

Who are the market leaders? Do you have any role models? Who are the most successful people? (Note: You can also include people in other industries who you admire and who might serve as role models.)

2. WHAT ARE THE CHARACTERISTICS OF THEIR PERSONAL BRAND?

For example, do they have a background story that everyone knows such as a turning point in their life? A signature phrase which everyone associates with that person (like Nike's Just Do It)? They could have a nickname that describes their brand, such as The Jerk (obviously not someone known for their compassion). They might even have a style that distinguishes them, such as always wearing a baseball hat and sandals. List the things that distinguish the people you listed in Step 1.

3. WHERE ARE THESE PEOPLE SEEN MOST OFTEN?

The places where you are seen most often are the places where you will have the biggest opportunities to build your personal brand. Where are your market leaders hanging out and promoting their brand? For example, do they do frequent seminars or speak at large conferences? Do you mostly see them on Facebook, Twitter, YouTube, LinkedIn or Google+? Their blog might be their biggest source of branding, or radio shows and advertisements. Do searches online to see where their names come up most often.

4. WHAT DO YOU ADMIRE MOST ABOUT THESE PEOPLE?

List out some of the characteristics that you would want to model or adapt to your own brand. For example, they might have a very personal way of communicating with their market in which they share their own pains and heartaches. Or they might have a very non-nonsense attitude and be known most for their guerilla-style, aggressive marketing tactics.

5. WHAT MAKES THEM UNIQUE?

How do they stand out from the rest of their industry and the world? Why are people so attracted to them? For example, do they meet a specific need that no one else does quite as well?

PART B: BRAINSTORM YOUR BRAND

1. WHO ARE THE PEOPLE YOU WANT TO APPEAL TO?

What is the target market for your personal brand? Is it primarily your customers? What are some of the characteristics of your market? For example, how old are they, are they more conservative or liberal, what other brands appeal to them? Talk to people in your target market and get a feel for what makes them tick. Hang out in forums where they are and listen in.

2. HOW DO OTHER PEOPLE SEE YOU NOW?

Ask some people who know you to describe the way they see you. What do they see as your strengths and weaknesses? Ask them what stands out most. If possible, do an anonymous questionnaire so that you can get honest feedback on the way you're seen now.

3. WHAT DO YOU WANT TO BE KNOWN AS? (E.G., THE PERSON WHO DOES X)

What are the key traits you want people to associate with you? For example, do you want to be known as the 'one-step marketing solutions' expert? Or, you might be known for '10-minute fitness results' or 'compassion-based coaching'. Brainstorm different possibilities based on the people you want to appeal to and your own strengths.

4. WHAT'S YOUR STORY?

Why do you do what you do? Write down some of the details of your background that have impacted your life. Is there a good story here that you can use as part of your brand? Whether it's an integral part of your brand or not, people will always want to know your story, so write a short description here.

5. WHAT'S YOUR STYLE? (E.G., CASUAL, PROFESSIONAL, GRUNGE, ETC.)

Write down the specifics of your current style. For example, do you tend to be more casual or professional? Are you more of a 60's throwback or a futuristic trend-setter? At the same time, note which of these style elements would most appeal to your target market.

6. THE SKILLS YOU WANT PEOPLE TO KNOW YOU HAVE (NATURAL, VALUABLE TALENTS)

Make a list of the skills your think are most important to the target market for your personal brand. These should include the natural talents that would provide the biggest impact and value for your market, such as a keen problem-solving ability.

7. SIGNATURE PHRASE OR VALUE PROPOSITION (JUST A FEW WORDS):

Do you have a signature phrase, value proposition or personal tag line? This would just be a few words. For example, the phrase "Elementary, my dear Watson" was Sherlock Holmes' signature catchphrase. One of Steve

Jobs' famous lines was, "Design is not just what it looks like and feels like. Design is how it works." You may not become as famous as Steve Jobs, but you should have a short phrase that describes you.

8. WHAT'S UNIQUE ABOUT YOU?

Finally, brainstorm all the different ways that you are unique from others in your market. What makes you stand out? If you're not sure, go back to what other people have said about you. You can also think about ways that you would like to be unique, even if you aren't right now.

PART C: DEFINE YOUR BRAND

1. YOUR STORY

What is your background story? What are some of the key events in your life that have made you who you are today? What are some of your key accomplishments or turning points in your life that you want to share with others?

2. PERSONAL VALUES

What are your most important personal values? Some sample values could be around seeing the best in all people, laughing about something every day, learning something new every day, striving for excellence, helping others be their best, etc. List your top 5.

3. PERSONAL BRAND STATEMENT (STATE WHO YOU ARE IN 1 SENTENCE)

Write out your unique, personal brand statement in one sentence or phrase. This should reflect the element that defines you as a person. For example, Tony Shepherd calls himself "The Hippy Marketer". Helen of Troy was, "The Face That Launched a Thousand Ships".

4. YOUR UNIQUE VALUE PROPOSITION (HOW YOU UNIQUELY PROVIDE VALUE TO YOUR MARKET)

Now write out a short sentence or two that describes how you provide value to your market in a unique way, different from others in your market. For example, Scott Tousignant's value proposition is, "I lead by example and encourage others to sculpt their body into a work of art while living their life to their fullest potential."

5. MISSION (WHY YOU DO WHAT YOU DO)

Your mission statement tells the world a little bit more about why you do what you do. It's not your story. Instead, it combines the elements of your personal values, brand, and value proposition all into one statement that describes your personal mission for your life. This doesn't have to be about your whole life. If you are developing a personal brand primarily around your business, then your personal mission statement can be focused more on those aspects.

CHAPTER 12

EXAMPLES OF FAMOUS PERSONAL BRANDS

These are all taken from Welcome, About, or other sections of these famous people's social profiles or blogs. Facebook and other social media sites are a great place to find out how people have defined their personal brands and the ways they connect on a personal level with their market.

OPRAH WINFREY

As the visionary and leader behind OWN: Oprah Winfrey Network and formerly the supervising producer and host of the top-rated, award-winning The Oprah Winfrey Show, Oprah has entertained and inspired millions of viewers to live their best lives. Her accomplishments as a global media leader and philanthropist have established her as one of the most respected and admired public figures today.

THE DALAI LAMA

His Holiness the Dalai Lama is the spiritual leader of the Tibetan people. His life is guided by three major commitments: the promotion of basic human values, the fostering of inter-religious harmony and the welfare of the Tibetan people.

BILL GATES

"I'm lucky enough to get to work with great people and focus on some of the toughest problems the world faces. Things like developing low-carbon and cheap energy sources, improving global health for the world's poorest, and working to find the best ways to improve our education system. I'll share notes on Facebook about great organizations, ideas, and events that are helping us make progress. Join us to get involved, learn more, or add your thoughts and perspectives. Thanks for stopping by. Bill Gates."

PRESIDENT BILL CLINTON

"After a career in public service, I'm enjoying life as a private citizen, continuing my work in the areas I can have an impact on through my foundation, including global health, economic development, climate change, and the childhood obesity epidemic in the United States. I started the Clinton Global Initiative to bring together global leaders in business, philanthropy, and civil society to take action on pressing global challenges. In my free time, I help out where I can to assist people affected by natural disasters including Hurricane Katrina and the 2004 tsunami in South Asia. When I'm not travelling, I enjoy spending time in my old farmhouse with my family in Chappaqua, New York."

ALICIA KEYS

Passionate about my work, in love with my family and dedicated to spreading light. It's contagious! ;-)

CHAPTER 13

The following are a few sample mission statements of people and companies. You can see how much they vary both in form and length. However, they are all similar in that they communicate the core mission that the person or company wants to achieve, reflected in both their values and results.

"Let the first act of every morning be to make the following resolve for the day:
I shall not fear anyone on earth.
I shall fear only God.
I shall not bear ill toward anyone.
I shall not submit to injustice from anyone.
I shall conquer untruth by truth.
And in resisting untruth, I shall put up with all suffering"
—Mahatma Gandhi.

"To inspire, lift and provide tools for change and growth of individuals and organizations throughout the world to significantly increase their performance capability in order to achieve worthwhile purposes through understanding and living principle-centered leadership."
—Stephen R. Covey

"Facebook's mission is to give people the power to share and make the world more open and connected."
—***Facebook***

"Mary Kay's mission is to enrich women's lives. We will do this in tangible ways, by offering quality products to consumers, financial opportunities to our independent sales force, and fulfilling careers to our employees.

We will also reach out to the heart and spirit of women, enabling personal growth and fulfillment for the women whose lives we touch.

We will carry out our mission in a spirit of caring, living the positive values on which our Company was built." (Note: there are several explanations of core values that follow this statement." These are Integrity, Enthusiasm, Praise, Leadership, Quality, Teamwork, Service, and Balance)—***Mary Kay***

We love Sixes—Our Six Rules

'The perfect job portal,' says Jobtardis founder Kumar Vuppala, "Change is Inevitable! Growth is Intentional. Get a Job you want, find a job you Love and earn what you deserve. If you need a job then you need jobtardis now".

As we keep looking towards the end of the sky, the following are our Six core principles we rely on

World's First Knowledge Auction Portal

1. Inspired by Stephen Hawkings's M-Theory "Physics leaves No Room for God" Jobtardis leaves no room for despair. Opportunity should be everything when you look for a job.

2. 'Success is not Hocus Pocus but 'Focus Focus'—Focus on all those innovations required for Jobseeker's, Recruiter's & Employer's.

3. Inspired by Gandhi's "A customer is the most important visitor on our premises, he is not dependent on us. We are dependent on him. He is not an interruption in our work. He is the purpose of it. He is not an outsider in our business. He is part of it. We are not doing him a favour by serving him. He is doing us a favour by giving us an opportunity to do so." Provide better customer support.

4. 80:20. 80% of the Jobtardis features are free such as job postings, resume uploads etc. We believe we can make money being the best.

5. Local people know what they need. So employ locals where possible.

6. We understand "WoW" means pleasant surprise which never stops!!

CHAPTER
14

JOBTARDIS
PERSONAL
BRANDING
PROCESS

STEP1:

Click on business with us, and then click Personnel
Branding (PB)

STEP2:

Click on start PB process

STEP3:

Jobseeker will complete a form and then click on submit
to go through the payment

STEP4:

Then Jobseeker will be redirected to payment page for
payment

STEP5:

If clicks on CC avenue it will go through payment page to
complete the payment process

"ONLINE REPUTATION MANAGEMENT"

By Kumar Vuppala

WHAT IS
ONLINE
REPUTATION
MANAGEMENT?

Everything you do or say these days has the potential to end up on the internet. The world of George Orwell's book 1984 is far more of a reality than any of us care to admit. However, it is possible to control the way we are seen on the internet, at least to some extent, through online reputation management (ORM). This involves monitoring what is said or seen about a person, brand or company on the web, encouraging the visibility of positive information, and suppressing any negatives.

MONITORING YOUR ONLINE REPUTATION

Basic internet reputation management begins with careful monitoring of all mentions on the internet. This means keeping careful track of search engine results for key terms like your name, address, company name or brand name. Anywhere that your name could show up is a place that needs to be monitored, not just Google. Remember that there are all the social networks and other search engines that can turn up different results.

THE BASICS OF BUILDING A GOOD REPUTATION ONLINE

As you go about building your reputation online, there are three basic keys to remember:

- **Regularly post positive content.** You need to make sure that the information you want the world to see is the content most visible on the internet. Regularly publishing your own content in a variety of places increases your visibility online and lets people know who you are and what values or messages you represent.

- **Keep your private information private.** The best defense against any future problems online is to keep private information hidden from public eyes. Either don't post that type of information online in the first place, or check your privacy settings everywhere to make sure you are limiting who can see your details.

- **Deal quickly with any negative information.** If you do find that there is something inappropriate, nasty or unflattering about you out on the web, you'll have to counteract it quickly. Don't let a piece of negative content get entrenched in people's minds or, even worse, go viral on the 'net.

BASICS OF DEALING WITH NEGATIVE INFORMATION

It is probably inevitable that something about you will show up somewhere at some time that you don't want anyone to see. Even if it is nothing particularly earth shattering, you may want to get rid of it or hide it from the world. Your main options will be to try to delete it or to push it down in the search engine results. By dominating the results with positive content, people are far less likely to see that one item that might embarrass you or cause the wrong impression.

It doesn't matter if you are an individual, a small business, or a large corporation. You still need to be concerned about your reputation on the web since everyone uses the internet to search for information. One negative review, comment or photo can affect you for years to come. Monitor your online reputation carefully and take control of how the world sees you.

CHAPTER
16

Controlling the way the world sees you is far easier than you might expect. By taking specific actions, you can manage your online reputation and create a consistent brand and image that bridges both the online and offline world. Crafting your message and delivering it to the world online is just another component of marketing. Some would argue that it is one of the most important business strategies you can use.

Use these 5 easy steps to start building your online reputation:

STEP 1: ASSESS YOUR CURRENT REPUTATION

Start by doing a thorough assessment of your current online reputation. Conduct searches for your full name, company name, brand or other key phrase in Google and other search engines. Focus primarily on the first 2 to 3 pages in Google's results. Put this information into a spreadsheet, listing the urls, position in results and your comments on what's in that url.

STEP 2: IDENTIFY CHANGES NEEDED

Look through each of the results on your spreadsheet and identify things you want to change. If there is anything negative, mark that as a priority to either delete or push down farther in the results. For results that contain key information you want to promote, mark those as items you want to move up in the rankings.

STEP 3: DETERMINE YOUR BEST ARENAS

Now look at what is showing up most in your results. Are your highest ranking urls from Twitter and Facebook? Maybe you show up more on a personal blog or in YouTube videos? If so, these are the types of places where you can publish more information about yourself and influence your reputation. At the same time, be sure to identify places where your target audience hangs out. You want to be seen and heard in the places that are most important for the type of reputation you want.

STEP 4: CREATE A REPUTATION STRATEGY

With information about your current reputation in hand, you can now craft a strategy for building the image you want. Focus on both the what and the where of your content and message. What type of content will serve you best? This might include blog posts, press releases, articles for directories, tweets, Facebook pages, or interviews. Then decide where this content will be published and how. This might include which social networks to focus on, high traffic blogs, news sites, your own blogs and video sites.

STEP 5: START BUILDING

The last step in building your online reputation is to follow through on your strategy and start creating your content and promoting it in the search engines and elsewhere with SEO techniques. Be sure to have your name, brand or company listed as the author on each item published, so that you get credit for it. As you create the content that will influence your reputation, continue to monitor search results online. Track your progress and keep tabs on all mentions of your name and key phrases so that you can quickly respond to anything negative.

Once you have assessed your current online reputation and have a firm strategy in place, you can easily take control and manage the way you are seen on the internet.

Keep your message consistent across all your content and you will be building a solid base that will help you both in your business and personal.

While you can certainly use different strategies to repair your image online, the best way to protect your online reputation is to be proactive. There are a number of measures you can take immediately that act as a defense against any negative information that could show up on the internet and harm your reputation.

TIP 1: CREATE PROFILES EVERYWHERE

Make sure that you have profiles in your own name, brand and company name in all the major social networks and bookmarking sites. These include Twitter, Facebook and Facebook Pages, LinkedIn, Google+, Yahoo and YouTube in particular. You should also consider creating profiles at MySpace, Naymz, and any new or rising social networks.

Don't forget about the major bookmarking services and Q&A sites also, such as StumbleUpon and Yahoo Answers.

TIP 2: CREATE EMAIL ADDRESSES

Make sure you have an email address in your name at gmail, at the minimum. Since this is such a widely used email service, you don't want to end up with someone creating a gmail account in your name and sending information that looks like it came from you personally. You should also consider creating free accounts at widely used services like Hotmail and Yahoo among others.

TIPS 3: REGISTER YOUR PERSONAL DOMAINS

Even if you're not planning to have your own blog, register your name as a domain so that someone else can't claim it. You can use a domain registry like GoDaddy.com or Namecheap.com to claim your domain. Set it up so that it automatically renews each year.

You could also set up personal blogs at Wordpress.com, Blogger.com, and Tumblr.com. Even if you only put up a single page with some public information about yourself, it's worth claiming blogs in your name at these major sites. Look around to see if there are other places or sites in your market where people are active. Then set up profiles under your name there as well.

TIP 4: CHECK YOUR PRIVACY SETTINGS

In any profiles or accounts that have privacy settings, check them carefully to make sure you are only allowing appropriate information to be seen publicly. For example, you can set your Facebook profile privacy to hide your personal details from anyone other than people you have added as friends. The same goes for sites like Picasa, where people might be able to find embarrassing photos that you'll regret later.

TIP 5: ACTIVELY ENGAGE AND PUBLISH

While you may have set up profiles and accounts in a variety of places online, that doesn't mean you have to be actively participating everywhere. Unless you have an army of helpers, that would be impossible. You need to be in the places where you want to be seen.

Pick the sites where your market is hanging out or where they go for information. Make sure you are engaging with people on a regular basis in those places, answering and asking questions. If there are specific types of sites where people in your market go for information, make sure you are regularly creating valuable content and publishing it there for everyone to read.

If you protect your online reputation from the start, you can control at least some of the information that appears on the web. By making it more difficult for people to hijack your name and by keeping up a steady flow of content, you give yourself a significant edge against the type of people who might want to harm your image.

HOW TO REPAIR
YOUR ONLINE
REPUTATION
WHEN THE
WORST
HAPPENS

No matter how many steps you take to build up and protect your image on the internet, there may come a time that it gets attacked. When your online reputation has been damaged by negative information, you'll need to act immediately to repair it and prevent it from becoming a long-lasting problem. You have a couple of options for damage control.

IDENTIFY THE SOURCE

Before you can do any reputation repair, you need to figure out where the negative information or attacks are coming from. Use the major search engines and social networks to see what is coming up in search results for your name. The sources will probably show up on the first 2 pages in Google, but you should also use a Twitter search or a tool like Social Mention to see where you are showing up in the social networks.

OPTION 1: REPAIR IT YOURSELF

If you need to go the route of repairing your online reputation yourself, the first step is to delete anything negative. If it is something under your own control, such as on a personal blog in a comment, then just go and delete it. Rather than waiting for Google to crawl your site and find the change, you can also use Google's Webmaster Tools to remove the url from search results. Check in their support sections if you need help with that.

If there is something negative on someone else's site, you can go to that site's webmaster and request that it be removed. You can also respond directly if there are comments or complaints about you on a site. Addressing the complaints in a polite and helpful way will actually serve to build up your image further.

Another important step to take is to counter any negative information with positive content. Start regularly posting articles, comments, blog posts, social updates and other information that can push the damaging content way down in the search results. Use whatever SEO tactics you know to promote your positive content to the first pages of Google.

As an extra note, watch carefully for other private data of yours showing up online. If something like your social security number or credit card number is posted, contact Google to have it removed. Search their help functions for the correct url for submitting your request.

OPTION 2: BRING IN THE HIRED GUNS

There are specialist Online Reputation Management services that you can pay to repair your reputation. Depending on the extent of damage, this can turn out to be a very expensive option. However, if you have the funds, it can be the most effective. You can also hire these types of firms to monitor your reputation on an ongoing basis.

Be careful about who you hire to repair and manage your reputation. There are plenty of dishonest or inexperienced consultants who can do more harm than good. Get a recommendation if possible.

Regardless of whether you do it yourself or through the help of a specialist, you need to act fast to repair your online reputation when it has been damaged. A single negative statement about you can go viral quickly and spread throughout the internet. By monitoring your reputation diligently, you can spot the damage before it becomes uncontrollable.

Search Engine Optimization (SEO) techniques are a key part of anyone's arsenal when managing online reputations. While traditional SEO focuses primarily on how to get your site to page 1, SEO for reputation management is focused on pushing negative information further back in search results and promoting positive information to the top. The actual tactics, however, are very similar.

CONTENT MARKETING

The most widespread SEO technique for building and managing your image is straight content. You can use both new and existing content to market yourself, just as you would use it to push a site up in a search result. If you can get your content onto high authority sites, then it is more likely to show up higher in the search engine results. The more eyes that see your content and your name in a positive light, the better.

While content marketing for traditional SEO is meant for driving traffic to a specific site, when it comes to reputation management it's a bit different. Your goal should be to have content that reflects your message dominating the first two pages or more of Google's search results. That's what people will see. So it doesn't matter whether it's showing up on your own site or someone else's.

USING SEO TO PROMOTE EXISTING CONTENT

The content that you already have online can be used to help build your reputation, but you'll need to use SEO to push it as high as possible in the search engines for visibility. The most effective way to do that? Pick the content that best represents you and your message. Then build backlinks and promote it on social networks to achieve better rankings.

When you start building more backlinks to your content, start by looking at where the nearest competitor's links are coming from. For each site above you in the search engine, use a tool like Majestic SEO to examine where they are getting inbound links. Then you can try to build links from the same places.

Other SEO techniques that you can use include building links from high authority sites. These include posting to places like the big social networks—Twitter, Facebook, YouTube, Google+, and LinkedIn. However, you should also look at the high traffic blogs in your market. Offer articles to these sites that related to both their audience and the content you want to rank.

USING SEO TO PROMOTE NEW CONTENT

Whenever you develop new content for building your reputation, you'll need to use both traditional SEO tactics and specific, targeted approaches.

For content that you are creating to widen the visibility of your brand, use the standard SEO that you would use for any marketing. Optimize each page for a specific keyword in the url, title, headers, image tags, and a couple times in the written content. Once it is published, promote it through bookmarking, social media, articles on directories, press releases, videos, and guest blogging. You shouldn't have to do anything particularly fancy.

However, you can also publish content that is specifically designed to counter anything negative that is already in the top search results. One way to do this is to identify the negative keywords that those sites have used that relate to you, such as the word "scam". Then create your own content that is optimized for that negative keyword, but which contains a message you want people to hear.

For example, say your product has been reported as a scam because someone didn't understand it. You can then publish an article discussing the common misconceptions about your product that might lead people to think it's a scam. Push that up in the search engines and you've both countered the negative and prevented some misunderstandings in the future.

You should also be sure to build links between not only the pages you have authored, but also ones that contain your name in a positive light. This will lend more authority to each link that you want to promote. Keep track of all your major sites to see their rankings and adjust your tactics accordingly.

While you may have thought of SEO in the past as a way to rank your money-making sites, it's clearly an invaluable tool for managing your internet reputation. It helps you to selectively promote content that represents your brand while countering anything negative.

WHAT TO LOOK
FOR WHEN
HIRING AN
ONLINE
REPUTATION
MANAGEMENT
SERVICE?

Online reputation management services and consultants have become more and more common as the internet gets increasingly complicated and difficult to navigate. For any business needing help with managing or repairing their image on the web, hiring an expert can be your answer. However, it can be a delicate and potentially dangerous situation when you put your own reputation in the hands of another. Follow a few basic guidelines when hiring.

WHAT IS THEIR EXPERIENCE LEVEL?

Investigate the consultants' level of experience in online reputation management by looking at how many years they have been in business. Did they just start this business last year or has it been longer? If they haven't been in reputation management long, what did they do before that? Don't write them off just for the number of years. It's possible that their prior experience is related.

You should also look at what types of jobs they have done in the past. How big were their clients? What was the nature of their assignments? How long did each job last or is it ongoing? How much work was straight SEO vs. new methods of social media management?

Another way to judge experience is through looking at how many clients they have had in the past and how many they are working with now. Do they have the staff or hours available to devote to you?

GET PROOF OF RESULTS

Always ask for recommendations and referrals directly from the company's clients. Just seeing a testimonial in print or video is not enough since these can always be fabricated. Make sure you hear comments directly from current or previous customers and ask those people what they were happy and unhappy with. There is always something that a person didn't like, so push to find out what it was. That way you can avoid having it happen to you.

Look at the service's own reputation online also. If they haven't done a good job promoting themselves on the web, how can they possibly do it for you? On the other hand, if they have achieved good results for their own brand, maybe they can replicate those methods with yours.

The company you're considering should have some stats for you to look at. If they have been keeping good records from previous clients, they should be able to show you what they achieved over a specific period of time. Depending on the other customer's situation, this can give you an idea of the level of difficulty they're used to dealing with.

ASK FOR A PROPOSAL

Before hiring anyone, always get a detailed proposal. They should outline their plan of action for managing or repairing your reputation, along with how long they estimate it will take. Naturally, you'll want to know what they'll charge. It might be a flat fee for repairing the damage from one problem. It might also be an hourly rate for ongoing reputation monitoring.

It's absolutely critical to ask what happens if they don't achieve the results promised. What do they guarantee in terms of rankings and time frame? What if they don't achieve those results? Do you get your money back?

While hiring a reputation management service is similar to hiring any professional consultant, you need to be extra cautious. Your reputation is on the line, so you need to be totally confident in their abilities. An inexperienced consultant can cause more harm than good, so make sure you do your research thoroughly and know what you'll be getting for your money.

CHAPTER
21

With all the places on the web where your name can appear, it can be difficult to keep track of your reputation. However, there are many tools available that can help individuals easily monitor and manage their own online reputations. Here are 10 of the top tools:

GOOGLE'S ME ON THE WEB

This is a free tool provided by Google that helps you set up alerts for monitoring mentions of your name or other information you provide. You will need a Google Profile set up in order to use it. It is then managed from your Google dashboard.

KNOWEM.COM

This service searches social networks, domain names and trademarks for your name/brand. You can also pay to have them set up profiles for you in all these places.

Social Mention

Social Mention is similar to Google Alerts, but it searches social media for mentions of your name rather than Google search results.

Trackur

This is a full service online tool that gives you a dashboard for monitoring and analyzing your online reputation in a wide variety of accounts. It is a paid tool that starts at $18 per month. It's great for small businesses in particular.

HootSuite

HootSuite is one of many tools available for managing multiple social accounts. While it was originally meant for Twitter, you can also manage multiple accounts on places like Facebook, Facebook Pages, Tumblr and Ping. fm. You then have the ability to post to any combination of these. There are both free and paid options.

Ping.fm

This is an online service for distributing tweets and feeds to multiple networks on the internet. It's also great for taking your RSS feeds from different sites and distributing them to many places.

Disqus

Disqus was designed to help you manage and display your comments across multiple networks, such as both blog and forum comments.

ATOMKEEP

Though it is currently in beta mode, Atomkeep gives you the ability to sync your profile across many networks. Every time you make a change in one profile, you can update all your other ones with the same information.

TWITTER SEARCH

The search function on Twitter lets you save searches for specific terms, giving you the ability to monitor the results of these searches.

RSS FEEDS FOR SEARCHES

Many sites give you the ability to create an RSS feed for a specific search, so you can monitor it from your feed reader. These include feeds for Yahoo alerts, Technorati searches, and Yahoo Answers searches.

With the wealth helpful tools available for managing and monitoring your online reputation, you can easily get stuck deciding which ones to use. Try picking the ones that will be easiest for you, that cover the networks you need to be on, and which do not overlap with each other. Be careful when using these tools since you don't want to post the same information more than once to the same place.

BONUSES

Source of results (eg Google, Yahoo, Bing, Twitter):
Keyword for search (eg your name, brand):

Date:

Position in Results	URL address	Type of site (news,blog, video,social, etc)	You Control? (yes/no)	Comments (eg, positive or negative about you, other content)
1				
2				
3				
4				
5				
6				
7				
8				
9				
10				

11				
12				
13				
14				
15				
16				
17				
18				
19				
20				
21				
22				
23				
24				
25				
26				
27				
28				
29				
30				
31				
32				
33				
34				
35				
36				
37				
38				
39				
40				
41				
42				
43				
44				

45				
46				
47				
48				
49				
50				

CHAPTER
23

Add your link after the "-" in the tweet and check the number of characters before pasting into Twitter or your scheduler.

Article	Tweets	Number of characters (max 100)
What is Online Reputation Management?	Building your online reputation isn't just about sending a message—	68
	How to be strategic when creating your online reputation—	59
	What is all this talk about online reputation management?—	60
	A quick primer on reputation management on the web—	53
	Why you need to keep close track of your online reputation—	61

To 10 Online Reputation Management Tools	What are the best tools for managing your reputation online?—	63
	Tools for managing your brand through social media and more—	62
	The most popular tools and services for online reputation management—	71
	Best tools for keeping track of your online reputation and brand—	67
	The tools that make it easier to manage your online reputation—	65

CHAPTER
24

JOBTARDIS PERSONAL BRANDING

As a jobseeker, your reputation is your most valuable career asset. Whether you're climbing the ladder at your current company or seeking a new job or seeking more daily rate or more monthly take-home, in today's recession and semi recession environment, you must proactively and continuously position yourself for success. Who are you? What type of person you are? Your credibility, visibility, personality, and personal style all make up your brand.

Know, Build and nurture your reputation index a.k.a personal brand and you'll make yourself a must-have, can't-fail jobseeker—and you'll do it without having to be someone you're not.

RepuIndex.com (Part of Jobtardis.com offering) provides you a free tool (SearchMe) to know your present reputation index or online reputation management. No need to state; your personal branding will help you not only survive, but thrive, in today's dynamic and ultracompetitive job market or workplace. WithRepuIndex.com tools, you can build your brand that enables you to differentiate yourself and stand out from your peers.

Use Repuindiex.com branding makes you as an indispensable, memorable, and unique professional. After all Success takes more than just hard work; brand yourself and watch your career soar.

CHAPTER
25

JOB OPPORTUNITIES

Today's everyone are experiencing the unanticipated negative consequences of their digital decisions—from lost job opportunities and denied college and graduate school admissions to full-blown national scandals. It also examines how technology is allowing students to bully one another in new and disturbing ways, and why students are often crueller online than in person.

We are at a critical point in technological history—where social media is beginning to have more impact on how we are perceived than our in-person interactions. Two areas of the Web 2.0 with greater synergy than Personal Branding (PB) and Online Reputation (ORM). In fact, the importance of Online Reputation Management (ORM) for Personal Branding is such that it is highly improbable you will be able to negotiate your way to the top in the often turbulent waters of the online medium unless you are well versed in the more basic and helpful ORM tools and techniques.

We live in a digital democracy, where all of us are exposed to both grounded and ungrounded criticism by friends, colleagues, customers, acquaintances and total strangers.

76

Moreover, none of us are immune to the highly damaging consequences that our own uncouth, half-baked online contributions and actions—which may range from posting the wrong picture or video to unwittingly saying the wrong thing at the wrong time and in the wrong channel—can bring on to us

You can have a clean record in Google or the other search engines—especially in that all-important first page, where 93% of searchers will stop—when someone looks up your name, product, or service and yet have your reputation in tatters in the social media. Even when the implementation of Google Social has made this scenario more unlikely, it is still the case that you're monitoring should not stop with the search engines. What is being said about you in the social networks? What sort of comments do your contributions generate and how well are they received? What is the general sentiment about you and how are your actions or your inaction contributing to it?

QUICK ADVICE FOR WHEN THINGS SEEM TO BE GOING WRONG

Jobtardis can offer some quick advice for when things seem to be going wrong:

NEVER IGNORE THOSE NEGATIVE MESSAGES

Never ignore negative messages, opinions and/or complaints. Make sure you go to the core of why you are generating a negative reaction and never hesitate to apologize if you have made a genuine mistake. It cost you job.

Don't Show Your Frustration Online

Whatever you do, don't lose your cool when you are under attack. Stake the high ground. Don't add fuel to the fire by venting your anger or your frustration, whether the criticism is justified or unjustified in your own eyes. And reprisals are a 'no-no' unless you want to risk the attack going viral.

Blocking someone is last resort

Blocking or kicking someone out of your social networks should always be the last resort. Remember they can continue the (negative) conversation about you elsewhere! Beware who are you and make sure that someone talks about negative in online.

Play Defense

Be prepared to stick to your guns and remain firm in the defense of your brand when your conciliatory efforts have failed and you are convinced that you've said or done nothing wrong. The sad fact is that we cannot please everyone no matter how irreproachable our actions. Some people are going to dislike you no matter what, but that does not give them the right to spread false rumors or make false allegations. Mount a strong defense and prove the opposition wrong when required.

Remain flexible

If your strategy and your approach are not giving you the desired outcome over a period of time, it may be time to re-think how you are putting your message across and to modify social media profiles, pictures, web design, and the nature of your online interactions. Remain committed to your values but be ready to change course and do not regard this as a defeat but as a lesson learned. Nobody ever said that building a successful brand would be anything other than a bumpy ride!

Online Reputation can truly help you in the pursuit of your Personal Branding and career goals. Jobtardis encourage you to adopt an Online Reputation Management strategy for your personal brand and invite you to become familiar with Online Reputation as one of the best guarantees of your online success.

CHAPTER
26

WHAT IS
REPUINDEX?

Repuindex is 'personnel branding' Site. Repuindex is a social media search platform that aggregates user generated content from across the web into a single stream of information. Repuindex provides tool to 'know' your brand reputation in online world. This simple site facilitates viewers to know their brand using our real time search engine page 'Search Me'. Our 'Brand Me 'Page enable users to check whether the desired brand is available online of not and then take them to services page. What is AIDA? Based on acronym AIDA—Attention, Information, Desire, Action How does it work? Based on Real Time Search Engine Results.

WHAT SOCIAL MEDIA PROPERTIES DOES IT SUPPORT?

Repuindex monitors 100+ social media properties directly including: Twitter, Facebook, FriendFeed, YouTube, Digg, Google etc.

WHAT SERVICES ARE PROVIDED?

Repuindex currently provides a point-in-time social media search and analysis service such as Build Branding, Monitor Branding for individuals and businesses.

WHAT IS "DESIRE"?

Desire is the likelihood that your brand is being discussed in social media. A very simple calculation is used: parse mentions within the last 24 hours divided by total possible mentions.

WHAT IS "ATTENTION"?

Attention is the ratio of mentions that are generally positive to those that are generally negative.

WHAT IS "INTEREST"?

Interest is a measure of the likelihood that individuals talking about your brand will do so repeatedly. For example, if you have a small group of very passionate advocates who talk about your products or brand all the time you will have a higher Passion score. Conversely if every mention is written by a different author you will have a lower score. Most frequently used keywords and number of times mentioned.

WHAT IS "ACTION"?

Action is a measure of the range of influence (jobability). It is the number of unique authors referencing your brand divided by the total number of mentions.

STEP1:

Click on business with us then click ORM

STEP2:

Click on start ORM process

STEP3:

Complete the form as shown below and clicks on submit and also complete the payment process

STEP4:

Complete the payment process. Upon completion client will receive a conformation email.

"SOCIAL MEDIA"

By Kumar Vuppala

THE TOP 10
REASONS YOU
SHOULD USE
SOCIAL MEDIA
TO PROMOTE
YOUR BRAND

Social media sites are networks where people interact with each other through personal profiles. They use them to hang out, keep in touch with friends, share interests, network and even find jobs. For your business, it's not a question of *if* but *when* you should start using them. Your business needs to be social and here are the 10 reasons why.

1. Everybody's There Already

Social media sites have hundreds of millions of users and a handful of them are among the most visited websites on earth. A large number of users check their profiles daily to see what's going on and they're growing in leaps and bounds. If you're there, you can be in daily contact with them.

2. Excellent Branding Opportunities

These sites put you in direct communication with your customers and there's simply nothing better for branding your business. Your customers can have a personal dialog with you.

3. LISTENING STATIONS

Social media also offer wonderful market research opportunities. Message boards, updates, groups and reviews allow you to find out exactly what your target market likes and doesn't like. You can use these resources to get to know them and better create products and campaigns.

4. LEAD GENERATION

You can use social media sites to get more customers. People will see your posts and hear about you from their friends. Plus, you get global reach 24/7 not only to your fans, but to their fans, and their fans, and their fans.

5. SEO BENEFITS

Profiles, updates and other social media content appear in search engine results pages and often rank quite high up. This is one of the biggest ways that people will find your profile. You also get backlinks coming into your site from your social media content.

6. THE PERSONAL TOUCH

Your social media presence and interactions with your fan base give your business a personal touch. That's what so many companies are lacking today. People like to know who they're buying from.

7. Easy Promotion

Social media sites offer the easiest way to reach everybody who might be interested in your sales and promotions. With one update, everyone you're connected to knows instantly when you have a special offer.

8. Rewarding Customer Loyalty

You can use social media sites to reward your fans' loyalty by offering exclusive discounts and freebies that are only available to them.

9. A Ready-Made Platform

There's no need to create a new website when you can simply make a profile. On most of these the social media sites, it takes only minutes to sign up and get started. Not only is it easy, but you also get the built-in SEO benefits of using their already popular website.

10. It's Free

You get advertising, marketing and branding tools at your fingertips without paying a dime for them. Social media networking is a dream comes true for businesses.

Before you get started, it's important to have clear goals. What do you want to do with your social media presence? What action do you want your fans to take? Stay focused on these goals and devote some time each day to building your social media presence.

Facebook is the king of social networking sites and it's not going to give up the crown anytime soon. There are over 250 million users worldwide and as many as half of them check their profile daily. Your business can use the site to reach out to any demographic worldwide.

Facebook offers fan pages for businesses. This is a mini-site for your brand where your customers can hang out and interact. Here are 10 tips for using your fan page to build your brand.

1. No Selling

First of all, the golden rule on Facebook is no selling. People use the site to socialize, find friends and share interests. Create a fan page where they can hang out and enjoy interactive features, and keep a lid on the sales talk.

2. Create a Vanity URL

When you first create your page, Facebook gives you a long, incomprehensible list of letters and numbers for your URL. Once you have 25 followers, you can make your own custom URL that's easier for folks to remember.

3. HOLD CONTESTS

Your fan page has a feature that lets you hold contests. You can set up a contest through your page and give prizes to the winners. This is a great way to get your customers involved in your brand.

4. MIND YOUR TABS

Tabs are the boxes that show up next to your business description. They include things like 'likes' and 'photos.' When people view your profile, only the first 4 are visible. The rest will be collapsed under a down arrow. Make sure that the ones you want people to see are at the top.

5. FEED YOUR POSTS

You can import an RSS feed to your fan page so that your blog posts will automatically appear there as a group of updates. This is not as complex as it sounds; just follow the prompts on your fan page.

6. YOUR ROW OF PHOTOS

The new Facebook layout has a row of photos at the top. Use these to show off your best products or services, or other images that help to brand your company.

7. OPTIMIZE YOUR 'ABOUT' PAGE

Your 'About' page contains your basic information and contact details. However, only a short sentence from the 'About' field shows up under your profile picture until people click on 'About'. Optimize this description with keywords and clickable links, and make sure to put the most important information up front.

8. WHERE TO SEND THEM

For better branding, you should use your page's 'info' URL instead of your wall. Your wall doesn't really tell them anything about your company and it may intimidate them to see posts by other followers.

9. CREATE CUSTOM TABS

Your page comes with standard tabs like 'info' and 'photos.' Delete tabs that aren't relevant to your business and create your own to replace them. Facebook has an app that lets you do this.

10. OFFER SOMETHING EXCLUSIVE

Offer special deals, free shipping or promotions that are available only to your Facebook fans. This gives them an incentive to become fans and also makes them feel valued by you.

Remember that whenever you're on Facebook, you should be 'on brand.' This may also be the case when you stay logged in or ask it to 'remember you,' even when you're surfing other websites. Because you're on-brand, don't say anything rude, negative or inappropriate. Be mindful of all that you do because it will reflect on your brand.

CHAPTER 30

10 WAYS TO BRAND YOUR BUSINESS USING TWITTER

Twitter is a social media and micro-blogging site where you send out short 'tweets' to all of your followers. With over 65 million tweets going out every day among its 160+ million users, your business needs to be there. Here are 10 ways to use Twitter to brand your business.

1. FOLLOW THE EXPERTS

Find experts in your field and follow their tweets. This shows your followers that you know what's going on in your industry and you can also learn how to market yourself on Twitter by watching what they do.

2. BE A SUBSCRIBER

Don't just subscribe to the experts, but also to anyone that's active in your industry. Twitter is the only social media site where users often follow more than they're followed. Follow lots of people and comment on their tweets.

3. Mark Favorites

When you find a tweet that you'd like to save for later, you can mark it as one of your favorites by clicking on the little star. Your followers can see your favorites listed on your profile. Your favorites list shows everyone a bit about who you are.

4. Ask Questions

Get discussions going on Twitter by asking questions. They don't have to be anything earth-shattering but can be about anything. Just throw them out there and let your followers respond. You not only get some activity going, but also learn valuable information about how your target market thinks.

5. Comment on Links

When you like something you see online, you can tweet it. It gives you the option to comment on your links and you should always do this. Just sharing without saying anything about it doesn't do anything for *your* branding; it only directs people to the link.

6. Make it Personal

People want to know about your brand but also they like to learn about you. Sometimes, it's good to tweet something purely personal as long as you don't do this too often. Twitter users love these tweets and they'll show the human side of your brand.

7. POST PICTURES

You can post pictures on Twitter and you should be doing this often. Choose pictures that are related to your brand and take your own photos showcasing your goods, your employees or the way your business works.

8. VALUE AND RELEVANCE

Whenever you get ready to post something, ask yourself these two questions: Is it relevant to my brand? Does it provide value to my followers? There are lots of tweets that don't deliver either relevance or value, so make sure yours do.

9. DON'T SELF-PROMOTE (TOO MUCH)

Once in a while, it's alright to show off your products and services, but don't be the type of user who shares it every time something good is said about your brand anywhere online. Focus on benefitting your users and not yourself.

10. MAKE THE MOST OF YOUR CHARACTERS

Since Twitter only allows you 140 characters on less in your tweet, you need to learn the art of concise writing. Check out magazine and newspaper headlines to see just how much you can pack into one short sentence. And make sure you use a url shortener so that you don't waste precious characters.

The most important thing to remember is that Twitter is supposed to be enjoyable. While you should be professional and realize that everything you do reflects on your brand, whenever possible offer something that's just plain fun. Show your sense of humor and be personable. Twitter is all about human interaction and that's what Twitter users want to see.

CHAPTER
31

**10 TIPS FOR
BUILDING YOUR
BRAND ON
LINKEDIN**

LinkedIn is a social networking site for professionals with over 120 million users. It's designed for making contacts, sharing ideas and generating leads. You can also use it to build your company's brand. Here are 10 easy ways to use LinkedIn to brand your business.

1. ANSWER QUESTIONS

LinkedIn has a questions section just like Yahoo Answers where you can answer question and share your expertise. Search for questions related to your business and provide helpful answers. This gets you credibility and exposure.

2. CHOOSE A SPOKESPERSON

Make one person in your company the LinkedIn spokesperson. It really helps to put a human face on your business. LinkedIn users like the personal touch. It also makes it easier for you to have just one person handling it.

3. UPDATE WITH PROMOTIONS

Whenever you update your profile, it sends a message to everyone you're connected with. This means that when

95

you add a promotion or special offer to your profile, the site automatically advertises it for you.

4. JOIN GROUPS

Search for LinkedIn groups that are related to your business. Choose a few that are active with lots of users to join. Commenting on groups is a great way to interact with people and spread awareness of your brand.

5. START A GROUP

You can also start a group for your brand. Offer exclusive incentives like freebies or discounts to group members only. Get a discussion going and find out how people feel about your products and services.

6. ADD A VIDEO

On Linkedin brand pages, there's a page called Products and Services where you can upload photos and videos. Instead of just uploading photos, make videos that show your products in action. Showing is much more powerful than telling.

7. TELL EVERYBODY OFFLINE

Use all of your offline promotional material to get people to your LinkedIn profile. Put your profile URL on all of your print materials and offer incentives if you want to grow your connections faster. Leveraging existing customers is a great way to boost your brand on social media.

8. Request Recommendations

'Recommendations' is a feature that gives you testimonials from people you're connected with. You can ask for recommendations from customers, clients, colleagues or anyone else you know. They boost your credibility by providing social proof.

9. Promote Your Events

If you have any offline events, promote them on LinkedIn even if you don't expect any of your connections to join. This shows that you're active in your community or industry, which is excellent for branding.

10. Combine Social Media Sites

When someone asks a question on Twitter, post it along with your answer on your LinkedIn profile as well. Do the same with Facebook and any other sites you're using. This shows your expertise and also provides you with fresh content.

The important thing to remember is that LinkedIn is for professionals. It's not widely used by everyone, or at least not yet. Keep all activity on the site strictly professional and focused on building relationships.

THE
BRAND-BUILDING
POWER OF
GOOGLE+—10
WAYS TO USE IT
EFFECTIVELY

Google+ is Google's own social media site. Fully integrated with other Google sites like Gmail and YouTube, it's still relatively new compared to Facebook, but offers great promise. It currently has over 20 million users and this number is growing every day. Creating a Google+ account for your business gives you an advantage in Google searches and also offers excellent features that help you build your brand. Here are 10 ways to use Google+ for brand-building.

1. WRITE A KILLER DESCRIPTION

Your short description is what will show up with your page in search engine results, so spend some time making it really good. It should grab attention, tell what your business is about, and be keyword optimized.

2. BE UNIQUE AND PERSONAL

Whenever you're interacting with people on Google+, be yourself and show your human side. Don't be afraid to use a little humor. It's important to be real and not appear impersonal.

3. USE SCRAPBOOK PHOTOS

Google+ lets you put photos at the top of your profile's main page. These are called 'Scrapbook Photos' and you can use them creatively to attract attention and brand your company. You can also change them as often as you'd like to.

4. TRACK TRENDS WITH SPARKS

The Sparks feature lets you search using keywords and find out what's trending. You can then post these trends on your Google+ page to show your fans what's going on in your industry. This helps to brand you as an expert who knows what's hot.

5. TELL THEM ABOUT YOU

Spend some extra time writing a great 'about' page that tells people who you are, what you do and your brand message. Go ahead and use your real name and link your 'about' page to your personal Google+ profile to put a human face on it.

6. GET RID OF USELESS TABS

Google+ gives you a whole bunch of tabs automatically and there are many you won't use. Delete the ones you don't use so that they don't clutter up your profile. When you click on 'Edit Profile,' you'll see where you can delete them.

7. THINK ABOUT YOUR POSTS

On Facebook or Twitter, it's perfectly acceptable to make crazy, off-the-cuff comments, but with Google+, you should be more careful. Currently, the site is used mostly for professional networking or as a news source. Spend some time on your posts and make sure they're relevant and appropriate.

8. SHARE EVERYTHING EVERYWHERE

If you use Google Chrome as your browser, you can download a program called Extend Share. This automatically gives you the option to share your posts to Facebook and all of your other social media sites, allowing you to brand across platforms easily.

9. HOLD A HANGOUT

Hangouts is a feature that lets you do live video chats or impromptu webinars. Pick a topic related to your industry and hold your own Hangout. This gives your fans a chance to interact one-on-one with you. Even the Dalai Lama and Desmond tutu have done a hangout!

10. DON'T BE GREEDY

Don't use Google+ only to promote your own brand. Devote some time to checking out other people's profiles and updates. Comment and interact with them. This boosts your brand visibility and attracts new fans.

Google+ is new but it is getting more active every day. The key to using it effectively is to set aside a little bit of time each day to interactions. There's a lot to learn and features are being added constantly. Always stay focused on this one question—What do you want to achieve with your Google+ presence? Stay focused and the time you spend on Google+ will be well worth it.

CHAPTER
33

**HOW TO USE
PINTEREST TO
BRAND YOUR
BUSINESS—10
TIPS**

Pinterest's website says that it's a 'virtual pinboard.' Many people call it 'Picture Twitter.' You create boards and pin pictures to them. These boards are fully customizable and you can share them with your friends.

Pinterest has new users every day and is now driving tons of traffic. It's new, but like all social media sites, it offers powerful branding opportunities for your business. Here are 10 tips on branding your business with Pinterest.

1. DON'T JUST PIN YOUR PRODUCTS

Don't only make pinboards for your products and services. Make boards for all things related to your business. For example, if you offer interior decorating, give them boards with colors schemes that are commonly used in the industry. Make boards with unique content your followers will like or find informative.

2. MAKE FRIENDS

Seek out popular Pinterest users and make friends with them. Post on their boards and they'll post on yours. The site is still new, so a little bit of networking can get you into the 'in crowd.'

3. Add a Pin Button

On your profile's 'Goodies' page, you'll find a Pin It button that you can download and put on your site. This allows other Pinterest users to automatically pin your content with just one click.

4. Choose Niche Categories

People find your board by searching for niche categories. Choose as many diverse categories relevant to your business as possible so that you'll show up in lots of searches.

5. Show Your Business Inside and Out

GE did wonders for its branding by creating a board devoted to how they create products from the factory floor up. If you create a board that showcases the process of doing what you do, this gives your customers a peek inside your business.

6. Engage Your Audience

Ask for opinions and comments on your boards. Pinterest allows you to reward the most useful answers with a discount or giveaway.

7. Crowdsource

Ask your customers to take a picture of themselves with your product and pin it. They'll tag you in the pin and you can then repin it on your VIP board. Several companies have used this branding strategy very successfully to gain more exposure.

8. DAILY THEMES

Create a theme and launch a new pin each day. This is a great branding strategy because it gets people to check back with you every day to see what you've posted.

9. ME + CONTRIBUTORS

This is a really nice feature that allows your followers to contribute to your boards. It's a wonderful way to get them involved directly with your business, which in turn helps with your branding.

10. NO SELF-PROMOTION

Avoid any kind of self-promotion. This will destroy your brand on Pinterest. It makes you look like you're just there to make money off of people, and Pinterest users aren't into that.

Before you get too active on Pinterest, keep in mind that the audience started out as young, adult women. Topics like recipes, interior design, clothing, and do-it-yourself crafts are huge. Topics that don't appeal to this demographic aren't, although this is gradually changing. Make sure that there's a market for your brand on Pinterest before you invest too much time.

CHAPTER
34

YouTube is more than just a website; it's the new TV. When the president of the United States uses a website to broadcast his or her State of the Union Address, you know that it's a force to be reckoned with. However, YouTube is better than TV because it's free and user-driven. This is an extremely valuable tool for spreading brand awareness online. Here are 10 easy ways to get exposure for your business on YouTube.

1. KEYWORD OPTIMIZE

YouTube is owned by Google and videos appear high in search results, so make sure that your tags and descriptions are keyword-optimized. You might want to also optimize the actual content of the video because Google may use voice recognition software in the future to index videos.

2. SHOW YOUR EXPERTISE

The key is not to sell but to show your expertise. Pick a problem related to your business that your viewers might have and give them a solution. If you're a plumber, offer a simple plumbing tip. A cleaning company can make videos on easy carpet cleaning. Show them something you know how to do that will benefit them.

3. SHOW YOUR PRODUCTS IN ACTION

Don't just show your products or services, but show them in action. If you sell barbecue grills, teach your viewers some great BBQ recipes. Videos that show how to do something using the product are the most effective in marketing.

4. SHORT VIDEOS WORK BEST

Videos don't have to be long. In fact, they shouldn't be. Focus on making videos that are just a few minutes long. If you've got a topic that stretches over 5 minutes, make two videos covering different aspects of it.

5. GIVE THEM A SNEAK PEEK

Make videos that show your business's inside operations. Viewers who already like your brand love to get a sneak peek at what goes on behind closed doors.

6. SOCIALIZE AND SUBSCRIBE

Remember that YouTube is a social media site. Subscribe to other channels, watch videos, comment and make friends. Other users will do the same for you and you'll create a network of similarly minded businesses that share leads.

7. CUSTOMIZE YOUR CHANNEL

Make use of all of the customization options that YouTube offers for your channel. Try to give your visitors the best user experience possible. Align themes and colors with your website and other marketing materials.

8. PRODUCTION SCHEDULE

It doesn't matter how often you produce videos, but you need to do it regularly. Decide how many you'll post per week and stick to it.

9. EMBED EVERYWHERE

Put your videos all over your other social media sites. Embedding videos is as simple as copying a code and pasting it. You'll get them seen by more people and you'll also get backlinks.

10. MAKE 'EM LAUGH

Everybody wants to have a viral video. Just one video that goes viral can get you exposure like you've never dreamed possible. Most viral vids get passed around because they're funny, so add a bit of humor to your videos.

Once you start seeing results from your YouTube branding efforts, you might consider upgrading your video equipment. To get started, all you need is a good web cam. But investing in gear such as lighting, a better sound system, a camera and editing software can help you produce better videos more easily. It's well worth the money to create higher quality videos than your competitors'.

THE TOP 10
TIPS FOR
SOCIAL MEDIA
OPTIMIZATION

Social media optimization is a huge buzzword in the online world today. Also known as SMO or social SEO, it's all about optimizing your social media presence for traffic. While some of the same rules apply, there are also some differences you need to be aware of. Here are 10 tips on how to optimize your social media profile.

1. Socialize

This may seem obvious, but it's so important it needs to be stated right here at the top. Every social media profile you create should start with spending some time checking out other people's profiles, making friends, and commenting. It's all about networking. Follow them first and they'll follow you.

2. Use Multimedia

Use plenty of pictures, videos and other multimedia. Social media sites are designed for this and they make it easy to upload media files. When you use these, you create a much more interesting profile and it helps with branding.

3. Share Me

Add share buttons for your social media sites on each of your blogs or websites. These are buttons like Twitter's Tweet This and Pinterest's Pin It. Integrate everything as much as possible so that you've got traffic going in every direction.

4. Consistent Content

Just like your website or blog, you need fresh content. With social media, this is even more important. Get into a schedule of regular posts and updates, but also make sure that you're offering value or something interesting.

5. Keyword Optimize

Everything in your profile should be keyword optimized, and especially all titles, tags and descriptions. Be careful not to stuff.

6. Complete Your Profile

The search engines generally ignore half-finished profiles. It's also bad for your branding if your profile isn't complete. Fill out everything before you start networking.

7. Watch Visitor Behavior

Use the analytic tools provided by social media sites and pay attention to what people are doing on your site. This will tell you what's working and what's not. If some features aren't being used, eliminate them to cut down on clutter.

8. SHARE EVERYTHING

Create content that's unique, helpful and interesting, and encourage your visitors to pass it around. Viral posts get you massive exposure.

9. CONSIDER TIMING

Pay attention to the timing of your posts and track results. You'll find that certain times of day get you a better response. Focus on those high-traffic times.

10. IT'S NOT ABOUT YOU

Don't only talk about yourself and your products. Focus on topics that are helpful to your readers and the problems that your products or services can solve for them.

On social media sites, it's all about being real. If you consistently provide good content and increase user activity on your profile, the search engines will love you. Your social media optimized profile will brand your business and drive new fans to your site.

CHAPTER
36

Social media sites offer a way for people to communicate online and interact with their favorite businesses. Sites like Facebook, Twitter and Google+ are among the most popular and most visited on the Web, and many of their users check theirs every day. Here are ten ways you can use social media to better connect with your fans and boost your business.

1. IMPROVE YOUR CUSTOMER SERVICE

These sites give your customers direct access to you. This gives you a chance to listen to what they have to say and learn how they feel about your business. By being responsive and offering them access, social media works wonders for your customer service.

2. MANAGE CUSTOMER COMPLAINTS

Social media sites let your customers leave comments on how you're doing. When they complain or criticize your business practices, you not only get valuable feedback, but also a chance to reply and repair your image.

3. SHARE REAL-LIFE CUSTOMER STORIES

Social proof is powerful in establishing your company's credibility, and social media gives you lots of opportunities to get real testimonials. Positive reviews from your customers can be seen instantly by anyone who accesses your profile or theirs.

4. GET YOUR EMPLOYEES INVOLVED

You can use your social media sites for idea development. Give your employees a platform where they can share their ideas. You can also use it to keep in touch, offering newsletters, updates and employee incentive programs.

5. BOOST CUSTOMER LOYALTY

You can use your profile to run your customer loyalty programs by offering incentives and rewards. Everything can be done through the site so you don't need to deal with any more punch cards.

6. ATTRACT NEW BUSINESS

Your social media presence gives you free advertising on the world's biggest message board. People can find your profile through Web searches and their friends' profiles. This is a great way to attract new customers.

7. GET TO KNOW YOUR MARKET

Social media makes it extremely easy to get to know your target market. You can find out exactly who enjoys your products and services, as well as find out what else they like. Companies have spent millions in marketing research to gather this information and all you need to do is invest a little time.

8. Create a Brand Message

Every company has a brand message, like Nike's 'Just Do It'. You can use your social media site to communicate this message to your fans and prospective customers.

9. Connect Everything

By integrating your social media sites with your blog or website, you can generate more traffic and leads. Widgets and apps allow you to link everything together for maximum exposure.

10. Use Multimedia

Most social media sites have multimedia applications. You can make a welcome video showing what your company is all about or offer a streaming podcast to your customers. These help by branding your company and giving it a more personal touch.

The most important thing to remember about social media is that people don't use these sites to shop. They use them to socialize, share common interests, and generally hang out. Make your page an interactive place where they can get involved and spend time. This keeps your brand in their mind and whenever they're ready to buy, they'll come to you.

CHAPTER 37

THE TOP 10 MISTAKES PEOPLE MAKE USING SOCIAL MEDIA FOR THEIR BUSINESS

Social media sites offer a great way to brand your business for free. Everybody uses sites like Facebook and Twitter, and many check them daily. But there are some deadly mistakes that can really cost you and even kill your social media campaign if you're not careful. Here are the top 10 worst mistakes you could make.

1. Shameless Self-Promotion

By far, the worst thing you can do is use your social media site as a sales page. People on these sites don't want to be marketed to. They want to socialize, network and share common interests.

2. Buying Your Friends

On Fiverr.com and other freelance sites, you'll find people selling friends for social media sites. It may be tempting to pad your profile with fans, but this is a deadly mistake. In most cases, the 'friends' aren't even real people. It'll be obvious to the real people who show up on your site.

3. ONE-WAY EXCHANGE

When people comment or ask questions on your page, you've got to respond. You need to reply to them and make them feel welcome. This shows that you're an approachable, responsive brand that cares what people think.

4. NOTHING TO DO

What makes a social media site popular is its interactivity. If yours just presents information about your brand and doesn't give them anything to do there, people won't stick around. Use the apps and widgets available to make your site interactive.

5. SOCIAL MEDIA NEGLECT

Maintaining a social media site is a never-ending job. You should devote a little time to it every day. The brands that are most active on social media do the best with them. Carve out a little time and stick it in your schedule.

6. FIGHTING WORDS

No matter what anyone says about you or your brand, never be negative or nasty to them. That only brands you as a jerk. Always be considerate, polite and friendly. Keep things positive even when you'd love to give someone a piece of your mind.

7. It's All About Me

If your page does nothing but talk about you, your products and your services, people won't be interested. You need to find other things that your fans would like and share them.

8. Wallflower

Don't wait for people to visit your profile. Get out there and socialize. Make friends, spend time on other profiles, and get to know people. You'll be surprised at what a difference this makes.

9. Inappropriate Comments

Whenever you're logged in, you need to be extremely self-aware. Everything you say or do reflects your brand. Show your personal side but don't get too personal. You want to look like a brand owner and this requires a little bit of professionalism.

10. All Over the Place

Stick to things that are of interest to your brand. It's okay to digress and get off topic once in a while to show your personal side, but for the most part stick to things your fans care about.

The absolute biggest mistake you can make with social media sites is ignoring them altogether. Experts predict that social media sites will become the search engines of tomorrow. To some extent they already are. Sign up to each new social media site that comes along because it might be the next great branding opportunity.

TOOLS AND RESOURCES

Keeping track of all the social media that you need to be active on can be incredibly stressful and time consuming. Not only do you need to interact and engage customers, but you also need to monitor your activity to see where you are getting the most bang for your buck. As a result, a number of social media management tools have emerged to save you time and energy. Here are 10 popular ones.

1. HOOTSUITE

Hootsuite is a social media dashboard which gives you the ability to share status updates, links and media across several different networks. You can also share an RSS feed to your networks and give more than one admin access. There are options for monitoring conversations, setting up different streams of feeds, tracking of performance and scheduling of tweets. Hootsuite has both free and paid options.

2. Buffer

This social media scheduling site also has free and paid
options. Buffer makes it easy to instantly add an article
or other link to be shared to different social networks at a
future date. A simple button on your browser will open a
window in which you can write a comment and add the
link/update to your buffer. You can specify the different
times of day to post the updates, or Buffer will publish
them based on their research-driven time frames. You can
also post and add updates to your buffer by sending them
to your own unique email address.

3. Crowdbooster

Crowdbooster is a newer social media dashboard which
gives you detailed overviews and stats on your different
social sites, such as Facebook pages and Twitter accounts.
By showing you the analytics of your latest activity
and followers, Crowdbooster helps you identify exactly
where you should be focusing your efforts for maximum
"crowd-boosting". It also gives you the ability to schedule
your updates.

4. Tweetdeck

This is primarily meant as a tool for managing your
Twitter accounts, but also has options for other accounts
such as Facebook. The main feature of Tweetdeck is the
ability to section your tweets into different columns so
that they are easier to monitor. For example, you can have
different columns based on lists, mentions, messages,
accounts, friends and searches.

5. GOOGLE ALERTS

As one of the most popular tools for monitoring your online reputation, Google Alerts is indispensable to many people. You simply specify a keyword or phrase and Google will send you an alert whenever it sees those words show up in its search engine results. For example, by setting alert for your own name or business, in quotes, you will be alerted either instantly or via email or Reader of the exact search result where your name is showing up.

6. GOOGLE READER

Reader is Google's app for subscribing to and reading RSS feeds. You can subscribe to any site's RSS feed, separate those feeds into different categories, and share content from the feeds to various social sites or via email. You can even post the content to your Google+ profile, which is not an option on most other social media apps.

7. SOCIAL OOMPH

Similar to Hootsuite, Social Oomph is a social media dashboard that helps you manage your tweets, social updates, friends and followers. One of the nice features is the ability to manage things like automatically following people that follow you and sending messages to new followers. There are both free and paid versions, with the paid version giving far more features for optimizing your follower base. You can even post to blogs from Social Oomph.

8. DISQUS

Disqus is a platform that helps you manage comments and promote social conversation on your website. You need to create an account on Disqus and then install it on pretty much any type of site, Wordpress included. Commenters also have to create an account on Disqus in order to submit a comment. One of the greatest features is the ability to view and monitor the comments you have made on any Disqus-enabled website all from your one online account.

9. COMMENTLUV PREMIUM

CommentLuv is another way to promote conversation on your site. It is a plugin which shows a commenter's latest blog post, as well as giving them the option to show their Twitter account and main keywords. There are other features of this paid version as well, but the primary attraction for commenters is the exposure and links they get from commenting.

10. SPROUTSOCIAL

Sproutsocial is one of the most complete social media management dashboards. It has all the features that allow you to schedule updates, monitor activity, view analytics, target the right customers, and involve other team members in your social media management. In addition, the mobile app gives far more functions than other platforms and you can manage things like Foursquare and your contacts' information. Sproutsocial is a paid service, with different levels of pricing based on your needs.

In order to create an effective social media strategy, you can use this worksheet to lay out your primary goals and implementation strategy for each site in which you intend to be active. Include the following information

Social Media Site: List each site where you know your target market is active.

Goal: What is your primary goal for engaging with your market on this site?

Content to post: What type of content will you publish in order to best engage your market?

Posting schedule: How often do you need to post content and interact with people on this site?

Who? Who in your business will be responsible for managing this site, including posting of content and interactions?

Social Media	Goal	Content to Post	Posting Schedule	Who?
Facebook				
Twitter				

Google+				
YouTube				
Pinterest				
LinkedIn				

CHAPTER

40

Note: You are allowed 140 characters, but should make your tweets less than 100 to allow for your short link and for retweets.

Put a link to your article or product at the end of each tweet, after the "-". Make sure you don't go over the character limit.

Article	Tweets	Number of characters (max 100)
The Top 10 Reasons You Should Use Social Media to Promote Your Brand	Are you missing the boat on social media for your business?—	62
	10 reasons you can't ignore engaging in social media in your business—	72
	Why is social media so important for your business? Add to these 10 reasons—	78
	Not sure if you really need to be on social media? Read this—	63
	Your business needs to be active on social media, no matter what business you're in—	86

Using Facebook to Build Your Brand—10 Tips	Check out these 10 tips for building your brand on Facebook—	62
	Your fan page is one of your online brand centers. Follow these tips for building it—	87
	Optimizing your fan page photos, plus other tips for promoting your brand on Facebook—	88
	Make the most of the brand-building opportunities on Facebook with these 10 tips—	83
	What are you doing to build your brand on your Facebook page? Here are 10 tips—	81
10 Ways to Brand Your Business Using Twitter	Start mastering Twitter for branding your business with these 10 tips—	72
	How many of these tricks are you using to build your brand with Twitter?—	75
	Make the most of Twitter for building your brand with these 10 tricks—	72
	10 tips for branding your business through your Twitter activity—	67
	Increase your brand awareness on Twitter with these 10 tips—	62

10 Tips for Building Your Brand on LinkedIn	Use these 10 tips to tap into the brand-building power of LinkedIn—	69
	What are you doing to build your brand on LinkedIn? Here are 10 great ways—	77
	What does your LinkedIn activity say about your brand? 10 tips—	65
	There's more to LinkedIn than a professional network. 10 tips for building your brand—	88
	10 easy ways to make the most of LinkedIn for building your business brand—	77
The Brand-Building Power of Google+—10 Ways to Use It Effectively	Don't miss out on the opportunities to build your brand on Google+. Check out these 10 tips—	94
	If your business isn't on Google+ yet, you need to be. Read these 10 tips—	76
	10 effective ways to use the power of Google+ to build your business brand—	77
	Google+ is here to stay, so use these 10 ways to build your brand there—	74
	A growing, powerful social network for building your brand. 10 ways to use Google+—	86

How to Use Pinterest to Brand Your Business—10 Tips	Did you know that Pinterest offers a unique opportunity to build your brand? Here are 10 tips—	96
	Pinterest is a hot social network and your business brand needs to be there—	78
	Check out these top tips for branding your business with Pinterest—	69
	It's not just about pretty pictures. Use these 10 tips to build your brand on Pinterest—	90
	Does your market hang out on Pinterest? 10 ways to build your brand through pinning—	86
Use YouTube to Brand Your Business—10 Easy Tips	Any business that's serious about building their brand is on YouTube. Are you?—	81
	Are you making the most of YouTube to build your brand? Check out these 10 tips—	82
	Access the traffic-generating engine of YouTube with these 10 brand-building tips—	84
	How are you using YouTube to build your brand? Try these 10 tips—	67
	This isn't Hollywood. YouTube is just as good for brand-building as entertainment—	84

The Top 10 Tips for Social Media Optimization	SEO for social media is different from Google. Check out these 10 tips—	73
	Do you know how to optimize your social media for getting the most traffic?—	78
	Follow these 10 tips for getting the maximum traffic from your social media activities—	89
	How to optimize your social media efforts for the biggest business impact—	76
	Social media for business requires more than a few likes. Use these 10 tips—	78
10 Creative Ways to Use Social Media for Your Business	Get creative with the social media activities in your business. Try these 10 ideas—	85
	Social media gives you the opportunity to truly engage your customers. 10 creative ideas—	91
	Here are 10 creative ways to use social media in your business. How are you using it?—	88
	Your customers are on social media, so you must be too. Try these 10 creative ways to engage—	95
	10 ways to creatively engage your customers and interact on social media—	75

The Top 10 Mistakes People Make Using Social Media for Their Business	Are you slowly killing all your social media profiles with any of these mistakes?	81
	These 10 social media mistakes could be damaging your online reputation—	74
	Don't sabotage the social media efforts of your business. 10 common mistakes—	79
	Many businesses are still making these social media mistakes. Don't be one of them—	85
	Ignoring social media is one big mistake. Read on to learn about 9 more to avoid—	83